Yamile Venticuaba Hernández
António Empanga Baptista

Integral Agro-productive Systems For Agricultural Development

Yamile Venticuaba Hernández
António Empanga Baptista

Integral Agro-productive Systems For Agricultural Development

Integral Agro-productive Systems For Agricultural Development In Namacunde, Cunene Province

ScienciaScripts

Imprint
Any brand names and product names mentioned in this book are subject to trademark, brand or patent protection and are trademarks or registered trademarks of their respective holders. The use of brand names, product names, common names, trade names, product descriptions etc. even without a particular marking in this work is in no way to be construed to mean that such names may be regarded as unrestricted in respect of trademark and brand protection legislation and could thus be used by anyone.

Cover image: www.ingimage.com

This book is a translation from the original published under ISBN 978-3-639-68539-8.

Publisher:
Sciencia Scripts
is a trademark of
Dodo Books Indian Ocean Ltd. and OmniScriptum S.R.L publishing group

120 High Road, East Finchley, London, N2 9ED, United Kingdom
Str. Armeneasca 28/1, office 1, Chisinau MD-2012, Republic of Moldova, Europe
Managing Directors: Ieva Konstantinova, Victoria Ursu
info@omniscriptum.com

Printed at: see last page
ISBN: 978-620-8-63613-5

INTEGRAL AGRO-PRODUCTIVE SYSTEMS FOR AGRICULTURAL DEVELOPMENT IN THE LOCALITY OF NAMACUNDE, CUNENE PROVINCE

AUTHORS

MSc. YAMILE VENTICUABA HERNÁNDEZ ENG. ANTÓNIO ENPANGA BAPTISTA

SUMMARY

The aim of this research was to develop actions for the integration of production systems used in family farming in the town of Namacunde. The techniques used to gather information were interviews, surveys and observation, taking into account the following indicators: agricultural activities carried out, the way in which agricultural activities are carried out, livestock activities practised and the integration of production systems. The data obtained identified shortcomings that lead us to realise that in Namacunde traditional agriculture is practised, with insufficient integrated management of agricultural and livestock systems, as well as the perception that peasants do not take advantage of the benefits offered by integrated farming systems based on the principles of agro-ecology. These results made it possible to formulate a proposal for the management of integrated agro-ecological gardens, which influence agricultural development in the town of Namacunde, Cunene province.

***Key words**: Integrated production systems, family farming, agro-ecological management.*

INDICE

I. INTRODUCTION

It is estimated that by 2030 demand for food production will increase by 35 per cent, water by 50 per cent and energy generation by 40 per cent. These increases are related to predictions of population growth, intensification of urbanisation and changes in consumer behaviour. These are major challenges to be overcome in a relatively short timeframe. The scale of these challenges incorporates a series of dynamic variables to be considered as a whole, such as climate change, involving the need to mitigate and adapt crops and production systems, changes in land use, food security, as well as the need to reconcile agricultural production with environmental conservation and preservation (Observatório ABC, 2015).

The demand for food, bioenergy and forestry products, as opposed to the need to reduce deforestation and mitigate greenhouse gas emissions, requires solutions that encourage socio-economic development without compromising the sustainability of natural resources. The solutions to these challenges largely depend on sustainable soil management (Lal, 2004).

Agricultural production systems face major challenges due to globalisation, and it is of great importance that these systems improve their management in order to adapt to the various changes in the environment and maintain a sustainable level of competitiveness (Aguilera et al., 2003). Southern Angola is one of the regions with very specific characteristics, from the point of view of climate, physiography, vegetation and hydrography. Amongst other factors, this region is influenced by the latitude and the cold Benguela Current, one of the effects of which is the Namib desert, as well as more scarce and irregular rainfall. With climate change, this irregularity is increasing, jeopardising the livelihoods of the communities (Ribeiro and Satiaca, 2021).

According to the Republic of Angola (2009), agriculture in Angola before independence was the main sector of the economy and the country was a major exporter of maize, cotton and especially coffee. Family farmers, especially in the central and northern regions, were the main producers and were well integrated into the market. During the war, as mentioned earlier, a large part of the population was forced to leave their land and stop production in order to take refuge in urban centres or other safer areas of the country. Since the end of hostilities in April 2002, some of the displaced people have gradually returned to their areas of origin and started investing in agriculture again.

The choice of production systems depends on the interests developed by each producer, according to the relative scarcity of available resources in order to maximise income per hectare (Dufumier, 2016); generating interactions between subsistence farming, cash crop production and other means of diversifying income (Katan et al., 2005). Farmers are generally not used to managing home gardens; on the contrary, they prefer monoculture because it generates greater economic income (Palacios and Barrientos, 2014).

According to República de Angola (2009), the conditions of services, communication routes and markets are still being rehabilitated. The majority of farmers practise traditional agriculture using hand tools to prepare the land and eliminate weeds, planting local seeds left over from the previous harvest. In the provinces of Huambo, Bié, the coastal area of Benguela and in the south, practically in the province of Huíla, many farmers use animal traction. Some of them use tractors for land preparation as well as hoes; others use fertilisers and improved seed varieties.

Peasant farms, on a smaller scale, have also been based on intensive production models, which has led to their transformation into highly artificial ecosystems in the process of degradation. Angola is making remarkable efforts to neutralise the problems generated by conventional agriculture. To this end, work is being done to redesign these systems into others of an agro-ecological nature, with production diversification and integration of its components, in which environmentally healthy, ecologically sustainable, economically viable and socially just agricultural production predominates. Sustainable agricultural systems are based on knowledge. The ability of farmers to succeed depends on public research and education, either to increase productivity or to be based on the knowledge paradigm, formulated to achieve long-term economic, environmental and social sustainability (Ikerd, 2014).

The Instituto Superior Politécnico Tundavala [ISPT] carried out a study on the practices and techniques in use in family farming and pastoralism in the provinces of Cunene, Huíla and Namibe, with the aim of strengthening resilience and production in a context of climate change, with particular attention to the adoption of sustainable technologies suited to the prevailing agro-climatic conditions. This research favoured cereal crops (maize, millet, The diagnosis was ultimately aimed at providing knowledge on the most prioritised aspects (Ribeiro and Satiaca, 2021). Ultimately, the diagnosis aimed to provide knowledge on the most prioritised aspects (Ribeiro and Satiaca, 2021). These include:

• Strengthening resilience and family agricultural production in the context of climate change in south-west Angola, where the main limitation is the availability of water;

• Characterisation of the livelihoods, experiences and practices of the region's populations in terms of agriculture and pastoralism, including the identification of constraints, in order to design strategies, techniques and productive practices capable of supporting climate change mitigation;

• Identifying good practices that can be disseminated through field schools and agricultural schools.

In this context, the function of the mines as a basic unit is revalued and the space local integration is created, with a view to promoting transformations and achieving sufficient output in the production process.

The action to modernise family farming contributes to improving the living conditions and income of rural producers by promoting the strengthening of agricultural and livestock production enterprises aimed at integrating markets into regional value chains through services and inputs, in order to consolidate technological modernisation, increase productivity and improve the income of families living in poverty and extreme poverty in rural settlements. By financially strengthening farming families, it is possible to achieve a goal of sustainability.

Due to the importance of this study in terms of improving peasant economies and the standard of living of the rural population, it is presented as:

Scientific problem. Insufficient integration of agricultural production systems at family farm level determines agro-productive results in the locality of Namacunde, Cunene Province.

Hypothesis. If actions to integrate agro-productive systems are used, then productive results will be better at the family farming level in the locality of Namacunde, Cunene Province.

General objective. To develop actions for the integration of production systems used in family farming in the locality of Namacunde.

Specific objectives

1. Theoretically ground the general elements of integrating agricultural and livestock production systems;
2. To diagnose the current situation of family farming production systems in the locality of Namacunde, Cunene Province;
3. Design actions for the management of integrated agro-ecological gardens, which influence agricultural development in the locality of Namacunde, Cunene Province.

II. REVIEW BIBLIOGRAPHY

2.1. General Systems

Systems theory is a way of thinking about the world, an approach to problem solving. Development of problems and models, which includes consideration of a complex series of events, or elements, as a whole (Palacios and Barrientos, 2014). According to Neto et al. (2016), a system is a set of reciprocally related units. Every system has one or a few purposes. Elements (or objects), as well as relationships, define a distribution that always tries to achieve a goal.A production system is a combination of different subsystems: cultivation, defined at the level of homogeneously farmed plots, with the same technologies and succession of plantations (in this sense several agricultural systems are distinguished within a production process); reproduction, defined at the level of herds or flocks of animals; the processing of agricultural products (cereal processing, cheese making, etc.); and non-agricultural activities (Apollin and Eberhart, 2013). The self-sufficiency system is known as family production, as they are the ones who make the most of their productive resources (Palacios and Barrientos, 2014). The objectives of rural development basically centre on 4 aspects: Improving economic, human, political and environmental protection capacities. With regard to economic aspects, this refers to improving income from agricultural activities, non-agricultural activities, improving industries and developing infrastructure (IDB, 2014). According to Villa (2011), development has sought to reduce poverty and improve the living conditions of the poorest population. Raising living standards includes better income, labour supply, access to education, culture and human values.

2.1.1. Agricultural systems. Concept and classification

Agricultural systems are systems that have an agricultural purpose; they are also ecosystems; yes, they have one or more living components. An agricultural system is an agricultural component, that is, an organism that drives towards a utilitarian purpose for man (Lopez, 2007). According to Hirakuri et al. (2012), it is defined : The agricultural system refers to the regional organisation of the various plant and animal production systems, which takes into account the peculiarities and similarities of these different systems. This organisation should allow for the construction of production models and arrangements that describe, in the most accurate way possible, the production systems. production predominant in the region. The production system is made up of all the cultivation or breeding systems within a rural property, defined on the basis of production factors (land, capital and labour) and interconnected by a management process. An integrated system is when cultivation/breeding systems for different purposes (agriculture or arable farming, livestock and forestry) are integrated with each other on the same farm, with the aim of maximising the use of the area, the means of production and also diversifying income.

According to Loaiza (2014), there are several reasons why it is important to classify agricultural systems. The first is that the number of different types of systems you can deal with any one is very small. Constructing a complete scheme for classifying an agricultural system is a labour-intensive task, and little progress has been made so far. However, when discussing agricultural biology, it is necessary to consider the most important classification procedures for such systems and be informed about the broad framework that would be useful.

Production systems

The agricultural production system is made up of production subsystems such as: the annual crop production subsystem and the perennial crop production subsystem. In this production system, the farm cannot ignore the human component or rational entity, the farmer and his family. It is the farmer who makes production, resource and energy use decisions (MINADER, 2020). The farm as a unit of operation can be seen as an agricultural production system in which resources, inputs, agents interact as transformers managed by the operator and under constraints of physical, biological, social and economic environments, to contribute to the overall agricultural product.

For these authors, any concept of an agricultural production system must include at least the following:

a) A purpose for which the system was designed;

b) Borders: some procedure to say what is inside and what is outside the system.

c) Context: the external environment in which the system operates.

d) Components: main components that are related in the formation of the system.

e) Interactions between components.

f) Resources: internal system components.

g) Contributions used by the system, but coming from outside it.

h) Main products or desired achievements.

i) Useful, albeit incidental, by-products.

According to Astier, Masera, and Galván (2008), the subsystem approach differs substantially from the arbitrary consideration of parts of the system. Firstly, it must be recognised that any part of a system can be considered as another system in itself, examined in the same way and form. Thus, even if a system is built to be tested in the same way, obtaining such a model would be on a different level, appropriate to the system being examined.

Integrated agricultural and livestock production systems

An integrated agricultural production system can be defined as the integration of cultivation or breeding structures for different purposes in the same area, linked by a management process, with the aim of maximising the use of land and means of production, as well as diversifying income. It is based on sustainability, which, when focussed on the agricultural sector, is associated with the evolution of the production system (Kluthcouski et al., 2000).

Crop-Livestock-Forest Integration is defined as a sustainable production strategy that integrates agricultural, livestock and forestry activities carried out in same area, whether in the form of cultivation, succession or rotation. It encompasses diversified production systems for the production of food, fibre and energy, in order to optimise the biological cycles of plants and animals, as well as inputs and their respective residues. Agrosilvopastoral organisation is the most complex of the integrated systems. The crop options amortise the cost of implementing the forestry and livestock components, which in turn generate income in the medium and long term, thus balancing the economic viability of the system (Machado, Balbino and Ceccon, 2015).

Agricultural production benefits from maintaining the original characteristics of ecosystems through the synergy that occurs in the integrated system. Ecosystems are geographical sites whose components are closely related to each other, in such a way that a disturbance in one of them implies a change in others, events that are generally transitive due to the ability to re-establish these balances. In nature, ecosystems have structural and functional elements that are in constant dynamic interaction (Ordóñez, 2005), which offers possibilities for recovery under natural conditions.

The author goes on to explain that when man intervenes with actions that imply profound changes that alter these reciprocal relationships between the components of the environment in which he is acting, it happens in an agro-ecosystem or altered ecosystem. This condition weakens stability and, as a result, restoring balance is more difficult and slower. Subsequent interventions such as changes in land use alter these relationships and disrupt the level of harmony achieved.

The same source summarises a series of concepts related to land use change and its ecological implications. It points out that every change in land use implies a disturbance to the ecosystem that modifies one of the three components to varying degrees, depending on the extent, intensity and frequency of the change. The close relationship between forests, native grasslands, soils and their evolution produces an intrinsic sustainability that allows them to persist in the face of extreme climatic changes (high and low temperatures, droughts, floods, etc.). These highly biodiverse and resilient ecosystems do not need external subsidies to maintain themselves (Waring and Schesing, 2016).

According to Santana et al. (2019), for example, biodiversity is favoured, as it contributes to the appearance of habitats, niches for pollinating insects, potential enemies of pests, diseases, improves the biochemical and physical aspects of the soil, due to the increase in organic matter. The forest component acts as a windbreak, maintaining soil humidity, reducing the demand for irrigation, conserving water resources, reducing erosion processes, as well as intensifying nutrient recycling, among other benefits.

According to Balbino et al. (2019), consolidating the forestry component is also a challenging proposition for rural producers. The main obstacles that make it difficult to adhere to this component are the producer's unfamiliarity with the implementation and cultivation methods, the difficulty in finding qualified labour, the market price for wood and, above all, the economic indicators.

Market analyses must maintain a close relationship with knowledge of the production system. Cattle farming and agriculture already have economic indicators that make it easier for producers to make decisions. With the inclusion of integration forests, other points must be considered in the analysis, such as the geographical position of the production, specialised management of the property and the best species to be cultivated, always considering the multiple purposes of the system (Queiroz et al., 2016). Among the benefits of this production method are water and soil conservation, the recovery of degraded areas, animal comfort due to the shading of the trees, forage quality and the mitigation of greenhouse gases (Almeida, 2015). Other benefits of Crop - Livestock - Forest integration are explained by Behling, (2014); Machado et al, (2015). These include the following topics:

• Increased quality or improvement of soil attributes, which leads tothe recovery of degraded areas and, consequently, increased grain production;
• Reducing the idleness of agricultural areas, diversifying production (grains, meat, milk and wood products) and stabilising income on rural properties;
• Breaking the cycle of diseases, reducing pests and insects, as well as weeds, which reduces the economic risk of the activity;
• Improving the environment, positively influencing animal performance, resulting from trees properly placed in pasture areas and adding income by afforesting pastures; and
• Increased soil organic matter, leading to a higher infiltration rate andwater storage in the soil, with less loss through surface run-off, greater carbon sequestration and mitigation of GHG emissions.

2.2. Main agricultural activities and livestock

There are many activities carried out on the ploughs that complement the achievement of high productivity. According to https://www.ejemplos.co/15-ejemplos-de-actividades- agropecuarias/#ixzz87jFlXhG (2018), the following are described:

• Growing grains, cereals and oilseeds. One of the commercial sectors that generates and contributes the most volume of merchandise worldwide is seeds, cereals and grains. Whether for food, to nourish other crops or to introduce improved seeds, not to mention wheat, rice and maize, the cornerstones of the diet of the five moderates, this sector of the industry is possibly the most robust in the agricultural area as a whole.
• Growing vegetables. Large-scale vegetable production is the main food injection available in urban and suburban markets all over the world. The

demand for them is great that they are often grown in artisanal and organic ways, avoiding the effects of pesticides and agro-toxins.

- Fruit crops. Usually linked to seasonal fruit, these sectors have large areas of cultivation in which production is massively developed. Depending on the distribution channels chosen, these fruits can be distributed through the following channels This includes being sold on lorries travelling through the streets, especially when they come from small farmers. A high percentage also goes to urban industries and manufacturers, who make elaborate desserts and non-perishable consumer items with them.
- Greenhouse and nursery crops. Usually on a smaller scale, as these are crops that don't require large tracts of land but apply the laws of intensive agriculture in limited spaces with high yields, to produce a wide variety of vegetables that supply local demand. Many of these smaller crops are organic, and unlike traditional crops, they can be grown within cities.
- Floristry. Growing flowers for personal consumption or to make bouquets and arrangements is also an important industry, especially in countries like Colombia and Mexico, where they contribute to a not inconsiderable sector of the local economy in various cities.
- Forestry. This refers to the care and cultivation of wild vegetation, in woods, hills or mountains, allowing the extraction of materials (wood, cork, rubber) through more or less considerable industrial intervention, without transforming the natural space into a farm or cultivation area. Many of the materials used in light manufacturing come from these types of plantations.
- Cattle. Undoubtedly the most popular and widespread livestock activity in human civilisation, whose origins go back to remote antiquity and whose importance in most Western cuisines is undoubted, not only for its contribution to meat, but also for dairy products and a whole culture of exploiting hides for clothing and utensils.
- Pigs. Pigs rank second in importance in Western livestock farming, as their

meat is generously incorporated into the various diets of the hemisphere, both in sausages, chops and various preparations that come from virtually the entire body of the animal. In addition, its exploitation is relatively economical, since at least in small-scale livestock farming, it has become accustomed to providing food scraps and organic waste.

• Poultry. Raising and slaughtering chickens is also an extremely central economic activity in the livestock sector. Their meat is almost universally appreciated, as are the products prepared from their eggs, which allow for high profitability for the producer. However, it has often been questioned because of the use of hormones and other genetic supplements that are unethical and largely distort the consumption of this white meat.

• Sheep and goats. Less widespread in comparison, and meanwhile popular in Arab countries, Europe and Argentinian Patagonia, lamb and sheep grazing have also had their place in rural development and in the collective imagination. The rearing and sacrifice of goats and sheep is also appreciated, although not as much as cattle and pigs.

• Other species of livestock. There are other species of livestock adapted to the diversity of regions inhabited by man, which can be used as a direct and indirect source of food and which would also be included in the agricultural item, however particular or exotic they may seem.

• Livestock support activities. The agricultural sector also includes activities specific to the industry, such as the production of animal feed, distribution, slaughter or other secondary forms of utilisation which, however, take place in rural areas or, at most, in intermediate segments of the production chain.

• Fish farming and fish farms. Depending on the legislation, this item can belong to the agricultural sector or to coastal fishing. However, the captive breeding of gastronomically valuable species such as trout takes place in a way that is not very similar to the coastal collection of marine species, and is therefore closer to the livestock sector than to fishing.

• Beekeeping and honey collection. The creation and maintenance of bee hives for the extraction and harvesting of various types of products is also a well-known item in the agricultural sector. This produces honey, royal jelly, wax, pollen, propolis and apitoxins, all of which are popularly consumed and even have pharmaceutical value. Since the 1980s, however, there has been an alarming decline in the number of bees worldwide, which has been extensively studied by experts in the field, given the importance of these insects in pollination.

2.3. Agriculture from conservation

According to Verhulst et al. (2015), conservation agriculture is a broader concept than conservation tillage, a system where at least 30 per cent of the surface is covered with residues from the previous crop after the next crop has been sown. In conservation agriculture, the emphasis is not only on the tillage component, but on a combination of the following three principles:

1. Reduction in tillage: The aim is to achieve a system with zero tillage (i.e. no tillage) but the system can involve controlled tillage seeding systems that in general do not disturb more than 20-25 % of the soil surface.

2. Retaining adequate levels of crop residues and soil surface cover: The aim is to retain enough residues on the soil to:

• Protect the soil from water and wind erosion;

• Reduce water run-off and evaporation;

• Improving water productivity; and

• Improving the physical, chemical and biological properties of the soil, associated with long-term sustainable productivity.

3. Use crop rotation: The aim is to use a diversified crop rotation to:

• To help moderate/mitigate possible problems with diseases, illnesses and pests;

- Utilise the beneficial effects of some crops on soil conditions and on the productivity of the next crop; and
- Providing farmers with economically viable options that minimise risk.

These principles of conservation agriculture are applicable to a wide variety of crop production systems, from conditions with low rainfall productivity to conditions with high irrigation productivity.

According to the same source, the application of conservation agriculture principles will be very different from one production system to another. It is necessary to identify management components such as pest and disease control tactics, nutrient management strategies and crop rotation, among others, through applied research with the active participation of farmers. For example, in conditions with gravity irrigation, a system of permanent beds with furrow irrigation.

2.4. Edaphoclimatic characteristics of Angola

Angola has an excellent resource base for agricultural development, as stated by MINADER (2004), who says that climatic conditions vary from semi-equatorial forests and humid tropical plains in the north and north-east to dry, temperate highlands on the Central Plateau and desert areas on the southern border with Namibia. Most of the country lies between 1,000 and 1,500 metres above sea level, and rainfall varies enormously from 1,500 to 2,000 mm per year in the highlands, and 100 to 1,000 mm in the lowlands.

It also confirms that the country's water reservoir is enormous, with 47 river basins, of which 26 are permanent rivers and the rest are intermittent, flowing only in the rainy season. The soils are variable, in the south-east they derive from Kalahari sands, and provide a small base for agriculture, while in the highlands soils of granite and gneiss formations predominate, making them the most important for agriculture. Most of the soils are highly acidic and are often affected by aluminium toxicity, requiring the use of lime. Ferritic and paraferralitic soils are spread across both the Central Plateau and the north-east of the country and are suitable for agriculture.

2.5. Traditional Angolan agriculture, perspectives

Ecological constraints (climate, soil, etc.) and people's response to these constraints in order to better adapt to the environment will result in the distribution of the population in Angola.

According to Simões (2019), with almost all of its population essentially rural, of different ethnicities and, as such, with diverse socio-economic structures, Angola cannot be attributed to a single rural world, but to several. These rural areas are naturally associated with different types of economy. Peoples in the south-east, with an economy geared towards hunting and gathering, with almost no agricultural production whatsoever; the agricultural economy of the peoples in the north, with leftovers for barter, but no livestock; the pastoral economy of the peoples in the coastal south and south; and finally, the agro-pastoral economy of the peoples in the centre.

On the basis of the issues already addressed in the previous sections, an attempt was made to summarise traditional Angolan agriculture. At this point, a brief comparison was made between the two sectors that characterise Angolan agriculture: the business sector and the traditional sector. Subsequently, the territory was divided into large agricultural regions, according to the population density of the areas and the type of predominant activity practised, agriculture or agro-pastoralism (INE, 2018). In an article published by Rocha Mora, and Romero (2016), it states that the biggest problem with poor agricultural production is that most rural communities lack food security, as they are unaware of technical aspects in the management of their ploughs, such as poor soil conservation practices (derived from low soil fertility, poor drainage); Deforestation, (irregular rainfall and soil erosion), cultural, traditional aspects of ploughs, lack of knowledge, technical assistance to plant other crops, lack of resources to buy inputs, lack of production alternatives and diversification in agricultural production systems, all these factors limit production, putting their food security at risk.

The available resources are not well utilised, as they are used indiscriminately and in an irrational and unsustainable way. Most rural communities are vulnerable due to poor agricultural production (MINADER, 2022).

According to CEIC/UCAN (2021), the agricultural production system is made up of two subsystems, the entrepreneurial production subsystem and the peasant production subsystem. The entrepreneurial production subsystem is a subsystem formed by a minority sector that owns large tracts of arable land, also has the ability to hire moderate management and have technical and credit assistance, according to their needs. In terms of marketing, it eliminates all intermediaries for inputs and products and most of its production is destined for export.

According to the same source, among the characteristics of the peasant production subsystem, its members make up a large mass of the population: predominantly farmers with small tracts of land, most of whom have no agricultural vocation. As for their production, most of it is destined for self-consumption or family subsistence. Although their financial and material resources are scarce or non-existent, they abuse them and exert unequal demographic control over natural resources, especially land. They also have little access to credit, very little or no available capital and insufficient or non-existent economic incentives, a low level of education, traditionalism and the absence of rural community organisations.

If the means of production and the labour power belong to the worker, the products of his labour belong to him. In simple commodity production,

production is basically for consumption and not for exchange. The limited development of the productive forces and the social division of labour mean that mercantile or peasant production only develops in an incipient, secondary and subordinate way (Republic of Angola, 2020).

2.6. Family farming at Angola

According to Prevedel (2023), family farming is carried out by families, with production geared towards their subsistence. This activity takes place on small properties, with cultivation done by labour necessarily family. Normally,a agriculture developed by families is more sustainable. In addition to these characteristics, family farming also has:

- Greater productive diversity;
- Low use of agricultural inputs;
- Low use of technology.

The definition of family farming for developed countries is different from that for developing countries, such as Angola, since in the former case this type of farming is integrated into the market and is dedicated to the production of tradable products. As such, the family farming referred to in this article is usually referred to as peasant or traditional farming. This term is used to identify the social class that lives in the countryside, from the countryside and for the countryside and that practises essentially subsistence farming, only selling surpluses on local markets. The inclusion of this social group is fundamental to characterising Angolan agriculture today.

For Neto (2008), five essential features characterise peasant society: its complete autonomy from the surrounding society, which dominates it but tolerates its originalities; the importance of the structural organisation of the family group in organising its economic and social life; despite this society's relationship with the surrounding society they live in a society that is self-sufficient in economic terms; relations with neighbouring collectivities are weak, with internal relations of inter-knowledge being much stronger; and those responsible for measuring between peasant collectivities and the surrounding society play a fundamental role. Avillez (2013) calls this type of agriculture traditional agriculture, defining it as agriculture that has reached a certain long-term equilibrium, both in terms of the utilisation of available factors of production and the investments needed to increase the supply of these factors.

In Angola, according to Pedro (2008), agriculture is fundamentally of the family type, where the use of advanced technology is scarce and manual tools are normally used for land preparation, weeding, sowing and harvesting. The use of agricultural inputs is low and yields are usually low when compared to world production. According to MINADER (2004), it is estimated that traditional (or peasant) farmers account for 80 per cent of national production. Of the remaining production, 18 per cent comes from medium-sized farmers, and only 2 per cent from commercial farmers.

2.7. Constraints of Angolan agriculture

According to CEIC/UCAN (2021), the promotion of agriculture is fundamental when it comes to establishing a new model of economic growth that is sustainable and, above all, less dependent on oil. The agricultural sector is important not only in terms of promoting self-sufficiency and food security, but also in terms of supplying raw materials for the manufacturing industry and creating jobs, one of the crucial problems in rural areas, with a view to diversifying the economy. For the agricultural sector to grow sustainably, agriculture needs to evolve beyond the mere subsistence level. These authors emphasise that the main problems (obstacles) should be identified and solutions sought. The majority of the cultivated area is the responsibility of family farms, with an average area per holding of 2.32 ha, using low-efficiency production techniques, being unproductive, making it difficult for small family producers to create production surpluses, having little incentive to produce beyond the family's subsistence levels.

The Ministry of Agriculture's report on the results of the 2018/19 marketing year states that the number of tractors is low and that they are only used in the business sector. The animal-drawn plough is used in both the family and business sectors, but with an average per holding of no more than 0.2 and 0.3 tools in the family and business sectors respectively. Some of the constraints on the development of Angolan agriculture are related to the low level of human capital and knowledge, the availability of financial resources and the lack of incentives for producers and technicians. The low level of human capital and knowledge involves aspects related to:

- Low yields due to the low level of technology and difficulties in accessing quality agricultural inputs in the quantities needed and at the right time, such as quality seeds suited to the region, tools and utensils suitable for working the land, animal and motorised traction, fertilisers, means of preventing and combating plant and animal pests and diseases;
- Lack of an effective rural extension service that can contribute to technological progress;
- Research is not very advanced, with a lack of investment in programmes and laboratories;
- Cultural barriers and a lack of knowledge on the part of farmers and technicians in technical, management, market and other terms, necessary for the projection of modern agriculture;
- Poor business management;
- High production costs associated with importing inputs;
- Inexperience in operating in a market economy system on the part of the various authors, including the governing body;
- With regard to the availability of financial resources, the main constraints

relate to;

- Poor condition of roads, especially secondary and tertiary roads, and bridges, which hinders not only access to markets for the disposal of products, constituting a major obstacle to the commercialisation of agricultural products, but also access to the market for inputs;
- Lack or deterioration of other infrastructures, such as facilities for the different services, product storage and seed conservation;
- The lack of incentives for producers, service providers and technicians has the following consequences;
- Rural exodus, with the result that new generations are increasingly disconnected from the land and unfamiliar with the techniques of working the land, making it difficult for them to return to the countryside;
- Growing feminisation of farming, due to the exodus of young people to urban centres;
- There are no rural financial services or they are very weak when they do exist;
- Possible land conflicts that could arise between family farmers and businessmen due to the effects of the latter's presence in unused areas;
- The population's low standard of living, which limits expectations and demand for consumer goods;
- Unattractive environment for business investment.

2.8. Agricultural prospects at Angola

There are measures and efforts being made to combat poverty and food insecurity in several African countries, including Angola. According to MINADER (2004), on 9 July 2002 the African Ministers of Agriculture approved the Detailed Programme for the Development of African Agriculture (PDDAA) in NEPAD (New Partnership for Africa's Development). This programme aims to restore agricultural growth, rural development and food security in the African region. This programme comprises five pillars:

1. Expansion of the area under sustainable land management systems and safe control of water resources;
2. Improving rural infrastructure and trade-related capacities for better access to markets;
3. Increasing food supply and reducing hunger;
4. Development of agricultural research, dissemination and adoption of technologies to support long-term productivity growth;
5. Sustainable development of livestock, fisheries and forest resources.

In the government's objectives to promote the integrated and sustainable socio-economic development of the agricultural sector set out in the 2025 programme. Among the various specific objectives defined, we can highlight the formulation of an agricultural policy, the development of agricultural research capacities, the implementation of a credit line to support associations, cooperatives, small and medium-sized producers, the rehabilitation of infrastructure to support productive activity, the promotion of trade in rural areas, integrated promotion of strategic sectors, the diversification of the economy and the promotion of public-private partnerships. Priority areas for intervention in terms of agricultural policy for the development of agriculture include the promotion of rural trade, technical assistance and agricultural extension services, agricultural credit and scientific research (CEIC/UCAN (2019).

III. MATERIALS AND METHODS

3.1. Location

The research was carried out between March and May 2023 in the town of Namacunde, one of Angola's southernmost municipalities and the second closest to the capital Ondjiva, which makes up the Cunene Province. Its seat triumphs over the main route that leads to the commune of Chiede and other towns, such as Santa-Clara, where the border between the People's Republic of Namibia and our country is located. It is bordered to the north by the municipality of Cuanhama, to the east by the municipality of Cuangar, to the south by the Republic of Namibia and to the west by the municipality of Ombadja.

3.2. Characterisation of the Municipality Namacunde

The municipality has certain economic potentials, which are shown in the table.

Table 1. Agricultural information for the Municipality of Namacunde

Agriculture (practised on a subsistence basis)		Livestock (Cattle breeding)		Vegetation (Preservation and exploitation)	
Type	Function	Type	Function	Type	Function
Massango	A source for diet of the population.	Cattle	Source of wealth, investment, savings and de production	Forest (Small groves of colophospermum mopape-omufiaty).	Used for Material of construction of housing, enclosure of farming, feeding cattle and various species of acacia trees species of acacia, charcoal and firewood production.
Massambala	A source fordiet of the population	Goat	Source for the population's diet	Sunflower and other species	Logging wood and coal.
Courgettes and Macunde Beans	A source for diet of the population	Pigs	Source for the population's diet		

Source. Ministry of Territorial Administration and State Reform (2010).

The climate is semi-arid, characterised by very irregular rainfall and frequent droughts, which makes natural systems intrinsically fragile. The regional moderado is structured by particular geomorphological units which are the chanas, the small rainwater basins, the relief forms that define the great natural and cultural value. The chanas are the main drainage system for the surface water that flows southwards as a result of the regional Cuvelai floods. The flood is the main guarantor of the surface water supply system, which gives rise to groundwater. The soils are generally poor, i.e. derived from rocks and lacking in nutrients.

3.3. Population and sample

The population was made up of 11 peasant families from the town of Namacunde, and the sample consisted of 100 per cent of the population, as they all met the selection criteria of owning land for food production, the condition of the ploughs and animal husbandry.

3.4. Methodology used

The research carried out is of the exploratory-descriptive type, since we have analysed these conditions, seeking to diagnose and situate the agricultural and livestock systems in relation to the real conditions of the Namacunde mines.

A non-experimental design was used as the empirical research method, with the use scientific observation and research to capture information at a single point in time, with the aim of describing variables and analysing their incidence and interrelationship at a given moment, using the criteria set out by Hernández and Baptista (2003). The first step was to provide a theoretical basis for the propositions put forward in this study by reviewing the literature on the subject. The information gathered from the literature review was used to construct a questionnaire used to collect the data (Appendix 1).

3.5. Design methodological

The design was structured in phases that chronologically and systemically met the specific objectives of the study. The research requires a holistic approach for its analysis, which integrates the productive aspects, allowing for the development of an efficient design for its management and future implementation.

Phase I. Diagnosis of farming and livestock conditions in the community of Namacunde

Visits were made to each of the properties, including informal and fairly flexible interviews and surveys to characterise the production systems in the community under study. The variables assessed were:

1. Agricultural activities carried out;

2. How agricultural activities are carried out;

3. Livestock activities carried out

4. Integration of production systems

Phase II. Elaboration of actions for the management of integral agro-ecological gardens, which influence agricultural development in the locality of Namacunde, Cunene province.

This proposal for actions to achieve the operation of an integral vegetable garden is working on the basis of an agricultural production model that benefits the rural community, the region's economy and the environment. They function as an expanding modern production system, combining traditional peasant knowledge with current agricultural technology. Leaving behind the traditional use of the land, monoculture as a form of exploitation generates costs as a consequence, causing soils to become compacted, lose nutritional conditions and become less productive. All of this is based on the principles set out by Macieira (2017).

3.6. Analyses and processing of information

The research results were analysed using tables and graphs to make them easier to understand. The information was analysed using the Excel and Word packages in the Windows 10 operating system.

IV. RESULTS AND DISCUSSION

4.1. Characterisation of location

The panorama of inequalities has to some extent conditioned women's economic affirmation, with 4 women being heads of household and 6 being men, despite the central role of women in carrying out different agricultural production activities, marketing products and looking after children. The sample under study has an average age 32. These figures indicate that the farmers are adults with responsibility for creating a source of income to support their families.

They have a low level of education, which indicates little knowledge of production processes and possible integration alternatives.

Others are lacking in irrigation possibilities because they depend on the rain and have had difficulties in sustaining their families because the rain has sometimes been delayed. Chimpacas are fundamentally their source of water consumption. 100 per cent of the sample confirmed that there is little agricultural production, since there is insufficient exploitation of other crops, such as potatoes, yams, beans, manioc and fruit. According to them, the products mentioned are not part of their culture, and this is due to soil and climate conditions. As for preparing the soil, all the work is done manually, depending on donkeys and cattle with a mouldboard plough. They have also had difficulties acquiring seeds, fertilisers and organic fertilisers.

4.2. Agricultural activities carried out in the gardens

The graph shows agricultural exploitation in the study area, where tuber, vegetable and cereal crops are the most recognised with 88, 75 and 63 % respectively; followed by grains and fodder with 38 %. We must emphasise that none of the producers mentioned fruit trees. There were numerous fruit trees typical of the region, such as: Nombé, Nonhande, Nondunga, Nongongo and Figo. The latter are much appreciated by the population for their useful fruit. This coincides with what Mendes de Araújo (2023) said The result confirms that the bulk of the region's population's diet is based on cereal reserves, particularly maize, millet and sorghum, which they manage to preserve and which serve as a food base for the whole year, combining with other seasonal crops.

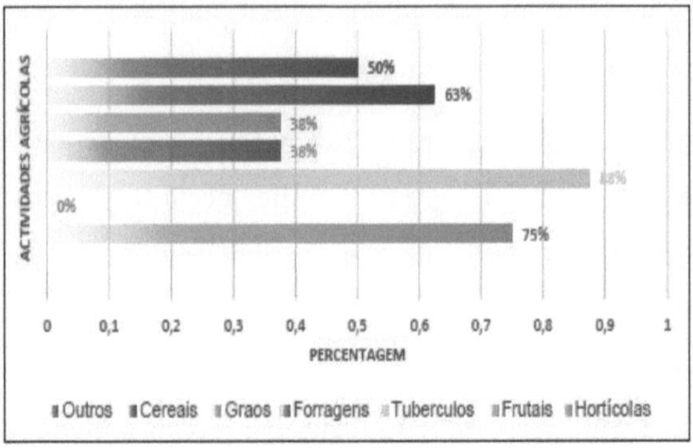

Figure 1. Agricultural activities most identified by producers

This can be compared with what the researcher Fernandes (2012) says when he states that the food consumption patterns of a domestic group must be perfectly coordinated in order to obtain the greatest calorific and economic return from its resources at each of the different points in the annual cycle, while always endeavouring to maintain the necessary food diversification throughout this cycle. Massango and massambala are crops found in 100 per cent of the lavras, which constitute the basic source of food for families in the region. There are around 50,000 species of plants on earth, 3,000 of which have been used for food, but only thirty crops are responsible for providing 95 per cent of the energy supplied by the food consumed by human beings; most of these (60 per cent) come down to rice, wheat, maize, millet and sorghum (Alisson, 2013). The levels of consumption vegetables must also be taken into account, Costa (2010) reports in his research that the average quantities consumed for vegetables and legumes were of reference, in the following percentages, for: Onions (29.1 per cent), tomatoes (27.9 per cent), carrots (33.3 per cent), lettuce (30.9 per cent).

According to these results, Ottmann et al. (2010) states that when a producer sows several crops in his garden, this practice brings social benefits, as the greater the search in productive and/or quantitative terms, it contributes to the expansion of production, the employment of labour and its consequent link to the market. For Capoulas (2017), in the context of the development of the population, especially the rural population, the promotion of family farming emerges as the basis of Angolan agriculture, fundamentally developed by peasants, who practise, still, systemsof agriculture using family labour, selling some of their surplus production to the consumer market, namely cassava, maize, beans, potatoes and also some vegetables and even fruit. This population is currently responsible for producing around 80% of cereals, 90% of roots and tubers and 90% of pulses and oilseeds.The same source says that the Massango and Massambala do Sul Zone is characterised as an arid area of the country, with frequent droughts and prolonged dry spells. These conditions lead to very variable grain production and, in general, lead to a lack and/or reduction of pasture and surface water, which can result in significant losses of livestock, which is a fundamental asset in the area.

4.3. Carrying out agricultural activities

Agricultural activities are not carried out in the same way by 100 % of producers; the majority, 39 %, do monoculture, 23 % polycultures, and the techniques of intercropping and crop rotation are carried out by only 23 % and 15 % respectively. This result shows that there is little knowledge of the benefits these practices offer to the soil and, in turn, to production results.

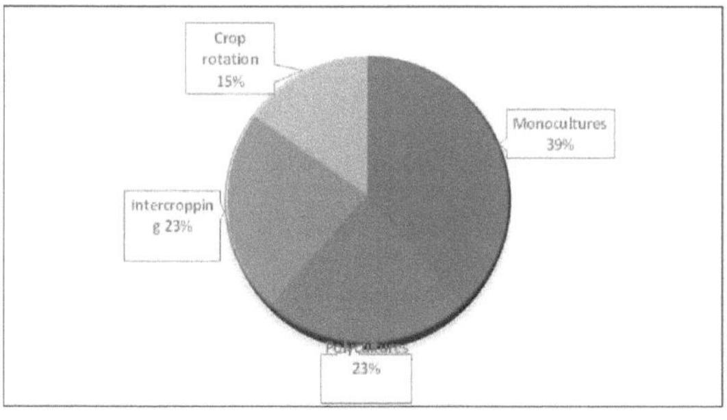

Figure 2. Agricultural activities and how they are carried out

According to Fernandes (2012), reasons such as ideology, religion, opinions, habits, moral conceptions, beliefs, traditional laws, etc. influence Angolans' culture of preserving the environment, which has increasingly deteriorated due to the constant intervention of different cultures in the country throughout history. The traditional practices of monoculture and post-cultivation he harvest is often practised as a belief and habit of thanking nature for the results of production.In this sense, Carvalho et al. (2016) recommend the use of intercropping to control pests and increase beneficial insect biodiversity, which influences plant pollination. They suggest using practices such as planting maize one year between the rows of guandu, which is more widely spaced, while in the following years guandu is planted as a single crop. After harvesting, the plants are cut down and left in the soil, without burning the residues. In the third year, maize is planted in the original guandu rows or in its ratoon. Another proposed rotation is to plant beans and maize in an area that was used to grow sweet potatoes the previous year; maize in succession to guandu and guandu in succession to beans of a different variety are also practical uses, all of which bring great benefits. addition, these results confirm what Freitas (2011) said, stating that the role of beans in restoring the fertility of the land is observed by farmers who grow them, and that there is great interest in cover crops that can be used as green manures.According to Capoulas (2017), family farmers have to make the most of their land so that they preferably have a variety of foods throughout the year. There are societies in which family farmers monoculture in order to exchange production for other foods. They produce several species at the same time, which they can change according to the season, if that makes sense; there are regions where climatic variables remain stable all year round. Many of them, not having the resources to buy improved seeds or agrochemicals from the industry, preferentially choose, save and sow seeds of plants that they consider most suitable in terms of production, resistance to disease or local soil and climate conditions. By doing this, they are practising a more "environmentally friendly" form of agriculture, protecting biodiversity, even without these express aims.

According to Hugo et al. (2019), several annual crops are used with crop-livestock integration in the Cerrado region, such as oats, millet, sorghum, maize, soya, cotton, beans, wheat and, more recently, sugar cane. But the most prominent are: Soya and maize, due to their area, production and economic importance.

4.4. Activity livestock

Table 2 shows the species of animals most exploited in the study area. Goats and oxen are the most numerous, with 38 and 31 per cent identified respectively, followed by birds in 25 % and pigs 7 %. These results are representative of the region, where ruminants are the most exploited species due to their adaptability to increasingly variable soil and climate conditions, and poultry and pigs are a minority due to the lack of food for them in competition with people's food.

Table 2. Livestock production of the study population

CAMPSIANS ANIMALS BY SPECIES					
	POULTRY	PIGS	CABRAS	BOIS	TOTAL
TOTAL	196	54	300	246	796
PERCENTAGE	25 %	7 %	38 %	31 %	

Although livestock is farmed, there is little production and the meat is not commercialised, as is the case in other parts of the world. The animals' food depends on nature and few producers keep fodder as crops. The properties visited have surplus areas of natural vegetation. As well as being located in areas that are difficult to exploit by agricultural activities, the farmers are aware of the importance of their preservation and regeneration.

These results are comparable to those of Lopes et al. (2018), who reveal that other widespread livestock farming are sheep and goat farming, with herds of 66,895 head of goats and 110,294 head of sheep, most of them raised extensively. Confined poultry shrank by around 50 per cent, while other herds grew, with other poultry up 450 per cent and pigs up 90 per cent.
%. The cattle, goat and sheep herds showed less variation than those, but were consistent.

In Brazil, as well as in other developing regions, animal production stands out for its importance in food security, job and income generation, draught power, transport, fertiliser and fibre production (Lima, 2006). Meat, eggs, milk and dairy products, together with plant products, can fulfil the nutritional requirements of human beings, they are sources of lipids, proteins and carbohydrates (Torres et al., 2000). These animal production systems are related to production capacity, management, social benefits and even the survival of families in situations such as periodic droughts. Thus, diversification of production is generally related to financial conditions, the personal characteristics of the producer and the factors of animal production in the economy of family farming, such as periods of water scarcity, soil acidity, distance from large input supply centres, among others (Bezerra et al., 2013).

4.5. Integration of systems from production

This indicator is shown in figure 3, where 100 % of the sample say they practice some kind of integration, and in all cases the livestock component is present, 27 % say they do it with trees, 55 % with horticultural species, only 45 % do it with cereals and tubers and 9 % use other integrations. This shows that producers, who still have little technical knowledge of the subject, apply empirical methods of integration, which to a certain extent are passed down from generation to generation, lacking a deeper understanding of the importance of agro-ecology and environmental conservation.

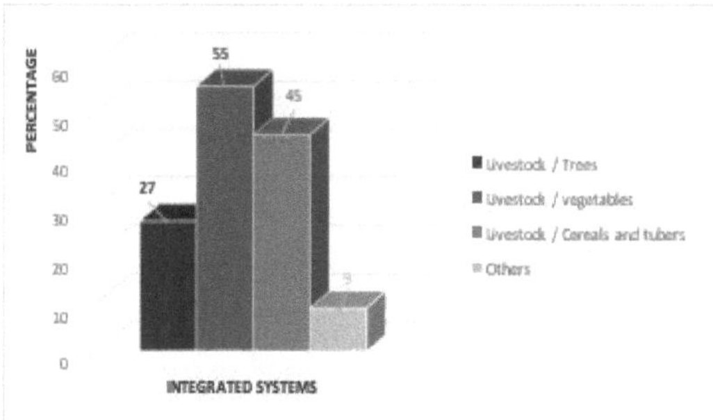

Figure 3: Integrated systems most used by producers

These results are comparable to those of Melo (2016), who states that soil protection measures are also known as "palliative measures" in ecological, biological, biodynamic or agro-ecological agriculture and constitute an alternative for improving agricultural soils, proposing an appropriate management of the natural resources that intervene in production processes. According to Hugo et al. (2019), the production systems used depend on the production focus and objectives of the property. On properties where the primary objective is to produce meat or milk, and where there is no interest in producing grain or conserved fodder due to a lack of equipment, the most commonly used annual crops are oats, millet or sorghum for grazing, and forage crops such as Panicums and Brachiaria can be established next to or after these crops. In more complex systems, the same crops can be used, as well as others such as maize and soya, which increase the fertility of the soil to higher levels and provide conditions for the production of hay, silage or grains for animal production or commercialisation.The excretions of grazing cattle contain considerable amounts of essential plant nutrients. This is supported by CAFS

(2022), when it gave the example of phosphorus (P), potassium (K), nitrogen (N), calcium (Ca), magnesium (Mg) and sulphur (S). These nutrients are returned to the pasture through the excretions of grazing animals. The value of feed as a source of nutrients depends on its distribution in the pasture, the animal category and species, the amount consumed and the chemical composition of the plant parts that are eaten. Only a small part of these nutrients are absorbed by the animals, with 60 to 90 per cent returning in the form of faeces and urine. For example, a cattle weighing 453kg produces 23.5kg of manure and 9.1kg of urine.The same source also confirms that when properly managed grain crops are rotated with pastures that are used correctly with animals, the soil's organic matter is renewed, the soil's physical structure improves and pore space increases. This production system is an important tool and can ensure the stability of farms. Lobato et al. (2017) state that natural products can be used to enhance the ecological management of agricultural agro-ecosystems. For example, referring to the management of rubbish (crop remains) in the backyard: "the bean straws are cut up and spread on the soil being prepared, then the 'fine' part of this partially decomposed residue is collected and taken to the vegetable garden beds. The "coarse" part is spread on other areas that are being opened up. This waste management is done only in the backyard area because of the quantity available and the distance from the other plantations."According to Dixon, Gulliver and Gibbon (2001) in mixed cereal-root crop production systems in semi-arid areas of Africa, the main source of In order to reduce this complex situation, it is necessary to use viable alternatives. In this regard, Arzola (2007) explains that adequate doses of organic fertilisers (animal manure) increase soil quality, improve humidity and protect the environment.

The diagnosis carried out covered the fundamental aspects where further work is needed, taking into account the characterisation of the agro-ecosystem's resources and in particular the study of the soils and water supply, from which the design of the property contributed to optimising the use of spaces according to their suitability and taking advantage of productive opportunities and potential, considering the synergisms of rotation and the association of crops. These results make it possible to create a proposal that could be viable for producers in the region.

4.6. Proposed actions for the management of integral agro-ecological gardens, which influence agricultural development in the locality of Namacunde, Cunene Province

Justification. Given its characteristics, the town of Namacunde needs a way of life for families settled in the countryside that as well as ensuring an abundant diet rich in proteins, vitamins and minerals (from meat, eggs, vegetables, fruit, cereals and milk), teaches each member to live in harmony with nature, preserving and enjoying their surroundings, breathing clean air, avoiding deforestation, conserving water sources and improving the land and crops.

The **Action Plan**

Action 1: Determine the areas for each of the garden spaces, bearing in that when the soils are fertile, smaller tracts of land can be used, and when they are not, a larger area is required.

Action 2: Determine the lines of agricultural and livestock production that are believed to be appropriate according to the conditions of the soil, water, climate and factors available in the region, and the market conditions, i.e. which products from the property are in demand on the market and offer margins of economic profitability to the producer.

Action 3: Locate the water sources for the vegetable garden; in addition, it is necessary to establish if water reserves are lacking, so that a well can be built.

Action 4 Draw up a map of the site, indicating the slope and the physical characteristics of the soils; take soil samples from the entire site and take them to the laboratory to be analysed. corresponding physico-chemical analysis.

Distribute the areas earmarked for fruit trees, vegetables, pastures, etc. on the map.

Action 5: Complement the different production activities with domesticated and wild species. Taking advantage of organic matter and energy to achieve high efficiency in the production process that works with less use of purchased inputs.

The proposal is based on a hypothetical area of 4 hectares of land, on which a family's agricultural, livestock and socio-economic activities will be carried out, with the respective integration of processes and under the principles of agro-ecology.

4 Socio-economic area and livestock	3 Pastoral area
1 Forestry and forage areas	2 Agricultural production area

Figure 4 - Graphical representation of the agro-ecological ploughing model for 4 ha.

Table 3 describes the key activities and functions to be carried out on each of the pieces of land belonging to the above proposal.

Table 3. Description of the agro-ecological ploughing model for 4 ha.

No.	Hectares (ha)	Characteristics and functions of the areas
1	1	Of trees with timber and fruit species (Nombé, Nonhande, Nondunga, Nongongo and fig), among them those that produce food (green mass), shade, fix nitrogen, organic matter and fodder species (Moringa, Leucena and Moreira), which also provide sufficient amounts of vegetable protein.
2	1	Used for sowing staple grains, root crops and vegetables (all native and improved species adapted to the region) with the application of organic fertilisers, intercropping, crop rotation, soil cover, wormwood culture, etc.
		For example After harvesting, keep the leaves covered and use the stalks as seed or animal feed. Cultivation in rotation and consortia of cereals and legumes and legumes and tubers Massango + guandu beans Maize+ courgette+ sweet potato Lettuce + beans + peppers Beans + massambala + cabbage Beans + sweet potato Beans+ aubergine+ tomato+ onion Ginguba + Corn Apply organic fertilisers, preferably manure from oxen and goats. Use repellent plants and bioplaguicides (mint and neem tree strata).
3	1,5	As a pastoral area, with the transformation of pastures and the creation of food sources for animals, in this case cattle, goats and other species. For example: Ensure that the area is divided into quarters with a hedge of fodder trees, preferably with a protein content and trees for shade and animal feed, green medicine and fruit trees are not excluded. We recommend leucena, mango for shade and neem for their rapid growth and foliage. Gradually replace natural pastures with species of higher yield and quality such as guinea likoni, brachiarias and star grass. Manage forage areas and establish king grass in order to carry out strategic grazing during the dry season. Sowing grain-producing species such as soya and massango and massambala for smaller cattle.
4	0,5	To build a , chicken coop, pig pen, ruminant corral, medicinal plant garden, stockpile seeds and surplus produce.

Expected results of the proposal

With the implementation of the integrated agricultural production system in the gardens in Namacunde, the biodiversity of the region will be improved and the sustainability of the families will be guaranteed. Not to mention improving soil and water quality, reducing the use of toxic agrochemicals and vulnerability to climate risks. From the point of view of livestock farming, the integrated system will improve animal welfare, due to comfort and management, ambience, increasing herd productivity. In agricultural activity there will be optimisation and intensification of soil nutrient recycling, improvement of its productive characteristics, diversification of business production, greater efficiency in the use of natural resources, new areas covered in vegetation, increased production of grains, fibres, organic matter, milk, meat, timber and non-timber products in the medium and long term.

V.CONCLUSIONS

1. In Namacunde, traditional agriculture is practised, with insufficient integrated management of agricultural and livestock systems.
2. Farmers don't take advantage of the benefits offered by integrated farming systems based on agro-ecological principles.
3. A proposal has been drawn up for actions to manage integral agro-ecological gardens, which influence agricultural development in the locality of Namacunde, Cunene province.

VI. RECOMMENDATIONS

1. Implementation of the proposal resulting from the research to the rest of the localities in the municipality, in order to promote the growth of productive indicators and the sustainability of families with farming operations.

VII. REFERENCES BIBLIOGRAPHIC

1. Aguilera, M., Bruna, G., Brzovic, F., Cerda, R., Clark, M., Chandía, A., Domínguez, J. I., Espinoza, A., Faúndez, M., García, P., Jara, C., De Kartzow, A., Kern, W., Lerdón, J., Marchant, R., Mora, M., Olavarría, J., Paillacar, R., Quijada, A., Troncoso, J. L. and Vargas,
G. (2003). Fundamentals of management for agricultural producers: topics and case studies agreed by Chilean universities. Programme. Santiago de Chile.
2. Alisson, E. (2013). Biodiversity loss is a global problem. FAPESP Agency, 2013. Available at: http://agencia.fapesp.br/17544.
3. Almeida, D. (2014). Urban Agriculture and Food Security in Belo Horizonte: cultivating a sustainable city. Agriculturas - Experiences in Agroecology, 1:25-28p.
4. Altieri, M. and Nicholls, C. (2013). Agroecología: Única esperanza para la soberanía alimentaria y la resiliencia socioecologica, 7(2), 65-83.
5. Apollin, F., and Eberhart, C. (2013). Analysing and diagnosing production systems in rural areas. Methodological guide. Ecuador: Quito, 237
6. Astier, M., Masera, O., and Galván, Y. (2008). Sustainability Assessment. A dynamic and multidimensional approach. Available at : http://docplayer. es/14885206- Evaluacion-de-sustentabilidad-un-enfoquedinamico-y-multidimensional.html.
7. Avillez, F. (2013). Lições de Economia Rural: Introdução ao Estudo da Economia da Agricultura Dita Tradicional. School year 1973/1974. Angola.
8. Inter-American Development Bank (IDB) (2014). Rural development policy. Inter-American Development Bank. Web Site: http://www.iadb. org/es/acerca-del-bid/politica- deesarrollo-rural,6229. Html.
9. Balbino, L. C., barcellos, A. O. and Stone, L. F. (2011). Crop-livestock-forest integration framework. Brasília: Embrapa, 127 p.
10. Behling, M. (2014). Crop-livestock-forest integration (ILPF). In:

FUNDAÇÃO MT. Soya research bulletin. Rondonópolis: MT Foundation, 2014. p. 306-325.

11. Capoulas, L. S. (2017). Family Farming and Sustainable Development in the CPLP. Minister of Agriculture, Forestry and Rural Development of Portuguese Government. Minister of Agriculture, Forestry and Rural Development of Portuguese Government.

12. Carvalho, A. H. de O., Delprete, S. I., Costa, A. C., Lima, W.L. and Oliveira, F. L. (2016). Agronomic and economic viability of lettuce and spring onion intercropping. Cadernos de Agroecologia, 11(2).

13. CEIC/UCAN (2019). ANGOLA ANNUAL ECONOMIC REPORT 2018. Catholic University of Angola, Luanda.

14. CEIC/UCAN (2020). ANGOLA ANNUAL ECONOMIC REPORT 2020, Catholic University of Angola, Luanda.

15. CEIC/UCAN (2021). ANGOLA ANNUAL ECONOMIC REPORT 2019, Catholic University of Angola, Luanda.

16. Costa, P. C. (2010). Characterisation of Eating Habits in Huíla Province, Angola. Dissertation submitted as a partial requirement for obtaining a Master's degree.

17. Sustainable Family Farming Course (CAFS) (2022). Technological degree course in family farming and sustainability integration systems FEDERAL UNIVERSITY OF SANTA MARIA CENTRE OF RURAL SCIENCES. 4th semester.

18. De Aquino, A. M. and De Assis, R.L. (2007). Organic agriculture in urban and peri-urban areas based on agroecology. Ambiente & Sociedade,1: 137-150p.

19. Dixon, J., Gulliver, A. and Gibbon, D. (2001). Sistemas de Producción Agropecuaria y Pobreza CÓMO MEJORAR LOS MEDIOS DE SUBSISTENCIA DE LOS PEQUEÑOS AGRICULTORES EN UN MUNDO

CAMBIANTE Principal Editor: M. Hall. FAO and World Bank, Rome and Washington, DC.

20. Dufumier, M. (2016). Importance of typology of agricultural production units in the diagnostic analysis of agricultural realities.

21. Fernandes, G. (2012). Livestock in family farming in south-east Angola, livelihoods and vulnerability of pastoralist and agropastoralist household groups. Technical University of Lisbon, Higher Institute of Agronomy. Lisbon.

22. Freitas, K. P. (2011). Production and appropriation of urban space in Campos dos Goytacazes - RJ: from single-family homes to flat blocks. Master's thesis- UENF Gliessman, S. R. (2000). Field and laboratory investigations in agroecology. Boca Raton: CRC Press LLC,. 330 p.

23. Hernández, S.R., Colinas, C. & Baptista, P. (2003). Research Methodology, 684pp.

24. Hirakuri, M. H. (2012). Production systems: concepts and definitions in the agricultural context. Documentos, Londrina, n. 335, 2012.

25. Hugo, Z.A., Motta, M. M. Cl., Volpe, E., Neivo, K. A. and Madureira, B. I. (2019). CROP MANAGEMENT IN CROP-LIVESTOCK INTEGRATION

26. INE, (2018). Inquérito Integrado sobre o Bem-estar da População - 2010, Ministry of Planning, Luanda.

27. Katan, C., Terenziani, F., Mongin, O., Werts, M., Porres, L., Pons, T., Mertz, J., Tretiak, S. and Blanchard, M. (2005). Effects of (Multi)branching of Dipolar Chromophores on Photophysical Properties and Two-Photon Absorption. The Journal of PHYSICAL CHEMISTRY A. Available at: https://doi. org/10.1021/jp044193e.

28. Kluthcouski, J. (2000). Santa Fé System - Embrapa technology: crop-livestock integration through the intercropping of annual crops with forage, in ploughed areas, in the no-till and conventional systems. Technical Circular, Santo Antônio de Goiás, n. 38, 28 p.

29. Lal, R. (2004). Soil Carbon Sequestration. Impacts on Global Climate Change and Food Security. Science 304: 1623-1627.

30. Lima, G.F. da C. (2006). Alternatives for the production and conservation of strategic forage resources in the northeastern semi-arid region. In: ENCONTRO NACIONAL DE PRODUÇÃO DE CAPRINOS E OVINOS, 1, 2006, Campina Grande. [Proceedings]. Campina Grande: SEDAP, 2006.

31. Lima, M. and Sidersky, P. (2000). Participatory diagnosis of native plants in four environmental moulds of Agreste da Paraíba. Recife: AS-PTA. 68 p. Work report for the "Native Plants Project".

32. Lobato, G. de J. M., Lucas, F. C. A., Tavares-Martins, A. C. C., Jardim, M. A. G. and Martorano L. G. (2017). Diversity of use and socio-environmental aspects of urban backyards in Abaetetuba, Pará, Brazil. Brazilian Journal of Agroecology, 12(2). ISSN 1980-9735.

33. Lopes, S. Y., Gamarra-Rojas, G., Paiva, F. F., Farias, J. L. de S. and Fernandes, C. de S. (2018). ANIMAL PRODUCTION IN THE FAMILY FARMING ECONOMY:
CASE STUDY IN THE BRAZILIAN SEMI-ARID REGION. Cadernos de Ciência & Tecnologia, Brasília, v. 35, n. 1, p. 53-74, jan./abr. 2018.

34. Lopez, R. C. (2007). Sistemas de produccion agricola sostenible en los Andes de Venezuela: Agricultura. Avances de quimica, 23-33. doi:1856-5301.

35. Machado, L. A. Z., Balbino, L. C. and Ceccon, G. (2015). Crop-livestock-forest integration: structuring crop-livestock integration systems. Documents, Dourados,n.110,Available at:
<http://ainfo.cnptia.embrapa.br/digital/bitstream/item/58600/1/DOC2011110.pdf.

36. Macieira, M. M. V. (2017). "Feasibility project for the implementation of an integral property in the Francisco do Orellana canton, for the year 2015" Thesis prior to obtaining the title of Commercial Engineer. SHOP - Ecuador.

37. Melo, L. P. (2016). The benefits of urban and peri-urban agriculture for the sustainability of the city of Macapá-AP. Pluris- CONGRESSO

LUSOBRASILEIRO para o Planeamento Urbano, Regional, Integrado e Sustentável.
38. Mendes de Araújo, E. S. (2023). Agricultural production systems on ploughs in the locality of Oncócua, Curoca Municipality, Cunene Province. End of course work in Agronomic Engineering. Ondjiva, 2023.
39. MINADER. (2004). National Medium-Term Investment Programme under the PDDAA NEPAD (PNIMP). Accessed on 13 May 2010, on the website : Ministry ofAgriculture e Ministry of Agriculture Development of Angola: http://www.minader.org/pdfs/fomento/programa_investimentos_medio_prazo.pd.
40. MINADER. (2020). 2019/20 Agricultural Campaign Report.
41. MINADER. (2022). Report on the 2020/21 Agricultural Campaign.
42. Ministry of Territorial Administration and State Reform (2010). General characterisation of Cunene Province. Available at http/infoangola.ao/index. Php.
43. Neto, J. (2008). Angola: Agriculture and Food. Portuguese Institute for Development Support, Documentation Centre. Lisbon, Portugal.
44. Neto, J. F.C., Baptista, F. O. and Cabral, C. (2016). "Angola: Agricultures and Food", in Angola - Agriculture, Natural Resources and Rural Development, Vol. I, Ilídio Moreira (ed.), Isa Press, Lisbon.
45. ABC Observatory, (2015). Low-carbon agriculture: the evolution of a new paradigm. paradigm. Available at: http://bibliotecadigital.fgv.br/dspace/bitstream/handle/10438/15353/Agricultura low-carbon agriculture: the evolution of a new paradigm.pdf?sequence=1&isAllowed=y
46. Ordóñez, M. J. (2005). Rapporteur of the sixth seminar on land use change. Retrieved 20 March 2005 from http://www.oikos.unam.mx/cus/index.html.
47. Ottmann, M. M. A., Borcioni, E., Mielke, E. and Da Cruz, M. J. R. (2010). Environmental and socio-economic impacts on community gardens under transmission lines in the Tatuquara neighbourhood, Curitiba, PR, Brazil. Revista Brasileira de Agroecologia, v.5 (1), p. 86- 94.

48. Palacios, V. and Barrientos, J. (2014). Technical and economic characterisation of production agrosystems in two indigenous reserves in Putumayo (Colombia).
49. Pedro, H. A. (2008). Public Irrigation in Angola: Proposal for a Management Model. Final Course Work in Agronomic Engineering. Instituto Superior de Agronomia Universidade Técnica de Lisboa, Lisbon. 50 pp.
50. Prevedel, D. (2023). Agricultural production: understand the types and systems of production. https://blog.aegro.com.br/producao-agricola/.
51. Queiroz, J.S., Gutierrez-Aleman, N. & León. F.A.P. (2016). The ecology and management of small ruminant production systems in the Sertão of Ceará, in the Northeast of Brazil. Agr. Syst., 22: 259-287.
52. Republic of Angola (2009). National Strategy for Food and Nutrition Security (ENSAN) Luanda, March 2009.
53. Republic of Angola (2020). Fundamental Report of the General State Budget (OGE 2020).
54. Ribeiro, C.S., and Satiaca, F.P.L. (2021). STRENGTHENING RESILIENCE AND FOOD AND NUTRITION SECURITY IN ANGOLA. Practices and techniques in use in family farming and pastoralism in the provinces of Cunene, Huíla and Namibe C.M.F 1 Instituto Superior Politécnico Tundavala, Tchitoco area, Huíla, Angola: Department of Engineering Correspondence: Carlos Manuel Fernandes Ribeiro Instituto Superior Politécnico Tundavala e-mail:cribeiroenator@gmail.com
55. Rocha, C., Mora, J. and Romero, J. (2016). Typology of production systems in the rural area of the municipality of Ibagué, Colombia. Agronomía Mesoamericana 27(2), 253-264. doi: 10.15517/am.v27i2.2436
56. Simões, J. (2019). Commercialisation as a factor of social promotion in Angola, Edition of the Commercialisation Fund, Luanda.
57. Torres, E.A.F.S., Campos, N.C., Duarte, M., Garbelotti, M.L., Philippi, S.T. and Minazzi-Rodrigues, R.S. (2000). Centesimal composition and calorific

value of foods animal origin. Food Science and Technology, v.20, p.145-150, 2000. DOI: 10.1590/S0101-20612000000200003.

58. Torres, E.A.F.S., Campos, N.C., Duarte, M., Garbelotti, M.L., Philippi, S.T. and Minazzi-Rodrigues, R.S. (2000). Centesimal composition and calorific value of foods animal origin. Food Science and Technology, v.20, p.145-150, 2000. DOI: 10.1590/S0101-20612000000200003

59. Verhulst, N., François, I., and Govaerts, B. (2015). Conservation agriculture: improving soil quality to achieve sustainable production systems? Centre International Maíz and Trigo Improvement Programme (CIMMYT) 2015. Available at (www.cimmyt.org).

60. Villa, I. M. R. (2011). What do we do with the countryside? Mundiprensa Mexico. Second Edition, pp 10-20.

61. Waring, R. H. and Schesing, W. (2016). Forest ecosystem concepts and management (pp.340). Academic Press Inc.

VIII ANNEXES

Annex 1. Peasant survey guide.

Dear producer, in order to characterise the production systems on your land, we are sending you the following survey. We hope that the criteria you provide will enable us to make proposals to improve the situation in the area.

Sex, Age, Educational level of the head of household.

Questionnaire

1. What agricultural activities do you carry out the most?

Growing cultivation , Vegetables ,fruit ,tubers ,fodder others

2. How are these activities carried out? Manual, mechanised ,with animal traction .

3. What livestock activities do you carry out?

Cattle farming, goats, pigs, poultry

4. How do you manage agricultural crops?

Monocultures , crop rotation , other

5. Do you integrate production systems at the mine? Yes No

What are the most common?

THE END...

Thank you very much for your attention.

Annex 2. Interview guide for a specialist from the Agricultural Company.

Dear expert, with the aim of characterising the production systems in the locality, we are conducting this interview. We hope that the criteria you provide will enable us to make proposals to improve the situation in the area.

1. What agricultural activities the local farmers carry out the most?
2. How are these activities carried out?
3. Is it easier for them to sow crops or raise animals of different species?
4. Which crops are sown the most?
5. Do you use any alternatives to improve the soil?
6. Which ones?
7. Do you integrate production systems in the mine? Yes No
8. What are the most common?

Annex 3. Livestock census of the families under study

CAMPAIGNERS	SPECIES AND QUANTITIES OF ANIMALS				
	POULTRY	PIGS	CABRAS	BOIS	TOTAL
1	15	8	8	14	45
2	5	9	0	21	35
3	17	5	28	25	75
4	13	0	35	0	48
5	12	3	0	60	75
6	15	8	50	0	73
7	35	0	32	65	132
8	23	0	40	0	63
9	17	6	35	7	65
10	21	13	37	32	103
11	23	2	35	22	82
TOTAL	196	54	300	246	796

Annex 4. Agricultural and livestock systems with signs of good practice in the locality of Namacunde

MIX
Papier aus verantwortungsvollen Quellen
Paper from responsible sources
FSC® C105338

Printed by Books on Demand GmbH, Norderstedt / Germany